# THIS BOOK IS THE PROPERTY OF:

**STATE** _____
**PROVINCE** _____
**COUNTY** _____
**PARISH** _____
**SCHOOL DISTRICT** _____
**OTHER** _____

Book No. _____
Enter information
in spaces
to the left as
instructed

| ISSUED TO | Year Used | CONDITION ||
|---|---|---|---|
| | | ISSUED | RETURNED |
| | | | |
| | | | |
| | | | |
| | | | |
| | | | |
| | | | |
| | | | |
| | | | |
| | | | |

**PUPILS** to whom this textbook is issued must not write on any page or mark any part of it in any way, consumable textbooks excepted.

1. Teachers should see that the pupil's name is clearly written in ink in the spaces above in every book issued.
2. The following terms should be used in recording the condition of the book: New; Good; Fair; Poor; Bad.

# Shop for Lunch

written by Barbara A. Donovan
illustrated by Michael Chesworth

Mom, Dad, and I go to a nearby mall. We want to shop for school things. We visit many stores.

At noon, we take a break. Lunch at the mall is such fun! It is not simple to make a choice!

First we try the fish shack. Will I order fish and chips? No, I won't get that today.

Then we visit Chester's Chop Shop. Dad's lunch is pork chops and applesauce.

We enter the sandwich shop. Mom orders a roast beef sandwich.

This is what I want for lunch! I'll get some cold cherry punch too.

We sit and eat lunch together.

Once we finish, we shop again. Mom and Dad haul bagfuls of school things. I have just one bag. It holds just what I wanted—a new lunchbox!